KU-189-236

Schools Library and Information Services
S00000639511

Rachael Bell

www.heinemann.co.uk
Visit our website to find out more information about Heinemann Library books.

To order:

Phone 44 (0) 1865 888066

Send a fax to 44 (0) 1865 314091

Visit the Heinemann Bookshop at www.heinemann.co.uk to browse our catalogue and order online.

First published in Great Britain by Heinemann Library,
Halley Court, Jordan Hill, Oxford OX2 8EJ,
a division of Reed Educational and Professional Publishing Ltd.
Heinemann is a registered trademark of Reed Educational and Professional Publishing Ltd.

OXFORD MELBOURNE AUCKLAND
JOHANNESBURG BLANTYRE GABORONE
IBADAN PORTSMOUTH (NH) USA CHICAGO

Designed by AMR
Originated by Ambassador Litho ltd
Printed by South China Printing in Hong Kong/China

ISBN 0 431 13333 6 (hardback)
05 04 03 02 01
10 9 8 7 6 5 4 3 2 1

ISBN 0 431 13338 7 (paperback)
05 04 03 02 01
10 9 8 7 6 5 4 3 2 1

British Library Cataloguing in Publication Data

Bell, Rachael, 1972–
Cows. – (Farm animals) (Take-off!)
1.Cows – Juvenile literature
I.Title
636.2

Acknowledgements

The publishers would like to thank the following for permission to reproduce photographs: Agripicture/Peter Dean pp 11, 15, 18, 19; J. Allan Cash pp 25, 27; Heather Angel pp 8, 26; Anthony Blake Photo Library pp 22/Maximillian, 23; Chris Honeywell p 29; Hutchison Library p 24; Images of Nature/FLPA pp 4/R. Bird, 7/L. Lee Rue, 10/Peter Dean, 12, 13 & 14, 17/Daphne Kinzler, 20/G. T. Andrewartha, 28/F. de Hooyer; Lynn M. Stone pp 6, 16; Tony Stone Images pp 5/Gary Vesta, 9/Dante Burn-Forin, 21/Andy Sacks.

Cover photograph reproduced with permission of Bruce Coleman.

Our thanks to Sue Graves and Hilda Reed for their advice and expertise in the preparation of this book.

Contents

Any words appearing in the text in bold, **like this**, are explained in the Glossary.

Cattle and how they look

Many farms keep cattle.

bull

Many farms keep cows for their milk and meat. 'Cow' is the word for the **female** animal and 'bull' is the **male** animal. 'Cattle' is the word for cows and bulls.

Cattle were first tamed in the Middle East about 8000 years ago.

It's easy to see why this is called a Longhorn!

Cows come in lots of colours and types. Many farmers keep black-and-white cows called Holsteins, because they make so much milk. Longhorn cattle are kept for their meat.

On the dairy farm

This farm has about 100 cows. Most of these are Holstein cows. They are mainly white with black markings. The farmer milks them twice every day.

Many dairy cows produce about 4500 litres of milk each year. But some cows produce up to 13,000 litres a year!

This maize will be turned into silage for the cows' winter feed.

On a **dairy** farm, half the land is grass, or **pasture**, for the cows to eat. The farmer also grows **maize**, wheat and barley. The maize is turned into **silage** for the cows to eat in winter. The wheat and barley are sold.

The cow and the bull

A cow produces milk to feed her calf.

The cow gives birth to one **calf** every year. The calf feeds from the milk in the cow's **udder**. When the calf is **weaned**, the farmer takes the milk that it would have drunk.

Bulls are bigger than cows.

A farmer keeps bulls so his cows will produce **calves**. Bulls are bigger than cows. Sometimes a bull has a ring in its nose so the farmer can lead it.

Cows give birth to calves about nine months after mating with a bull.

Calves

These calves are old enough to drink milk from the machine.

Baby cattle are called **calves**. A **calf** stays with its mother for the first three or four days to get the best of her milk. Then it drinks milk from a machine.

The calf has to have a tag put on each ear.

When the calves are a few hours old, the farmer puts a **tag** on each ear. The numbers on the tags say which farm the calf is from, and which calf it is.

A calf can stand up and feed from its mother soon after it is born.

Where are cows kept?

Cows stay outside on the pasture during warmer weather.

From spring to autumn, cows stay outside on **pasture**. The farmer moves them into a new field when they have **grazed** the grass short.

Most cows are kept inside during the winter.

In the winter, most cows are kept under cover to protect them from the cold. Every day the farmer puts fresh **straw** in the **yard**. He scrapes it out every three weeks.

What do cows eat?

The cow uses her tongue to pull the grass into her mouth.

Cows eat grass. The cow pulls grass into her mouth with her tongue. She chews and swallows it. Later she **chews the cud**.

A cow's stomach is made up of four parts, or chambers.

Cows eat silage during the winter.

In the summer, cows **graze** on fresh grass. In the winter, the farmer feeds them with grass or **maize silage** and a **mineral lick**.

Cows need to drink lots of water. They drink about 60 litres of water a day.

Why do cows live in groups?

Cows like to live in a herd.

Cows like to be together in a **herd**. They know which one goes first in the queue for milking and feeding. The farmer always keeps them in this order.

A big herd can look scary, but cows are really quite timid animals.

The mother cow licks her **calf** to rub off any insects.

Because cows live in groups, they can lick each other, to rub off insects. They also flick flies off with their tails and like to scratch on a post or a tree.

How do cows sleep?

Cows like to sleep on their sides.

Cows lie down on their sides to sleep. If they are outside, they choose a sheltered place and lie with their backs to the wind.

pasture

Cows rest for a total of eight hours out of every 24 hours.

Each day, cows spend about eight hours eating, eight hours **chewing the cud** and eight hours resting. In the daytime, they do each of these for only about 20 minutes at a time. How many hours do you sleep?

Who looks after cows?

This cow is being milked by hand.

Some people still milk their cows by hand. On other farms, the farmer and the **farm-hand** use a machine to milk several cows at a time.

The place where cows are milked is called a milking shed or parlour.

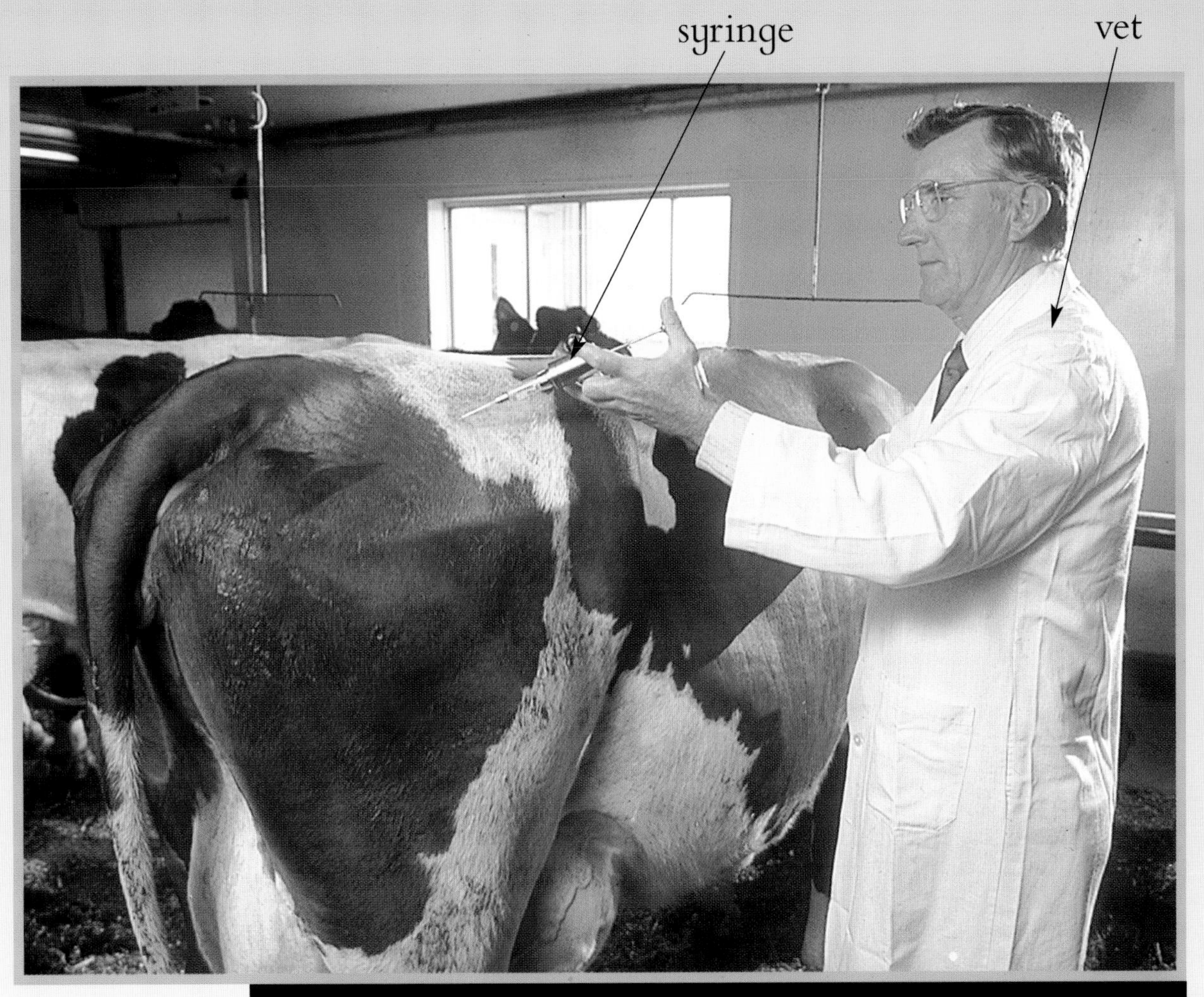

The vet is giving the cow an injection with a syringe.

The farmer and farm-hand clean the **yard** and milking shed. In winter, they also feed and **bed** the cows in the covered yard. If a cow is very ill, the **vet** is called.

What are cattle kept for?

All these foods are made from milk.

The cows on **dairy** farms are kept for their milk. Each cow produces about 4500 litres each year. The milk is bottled for us to drink, or made into yoghurt, cheese and butter.

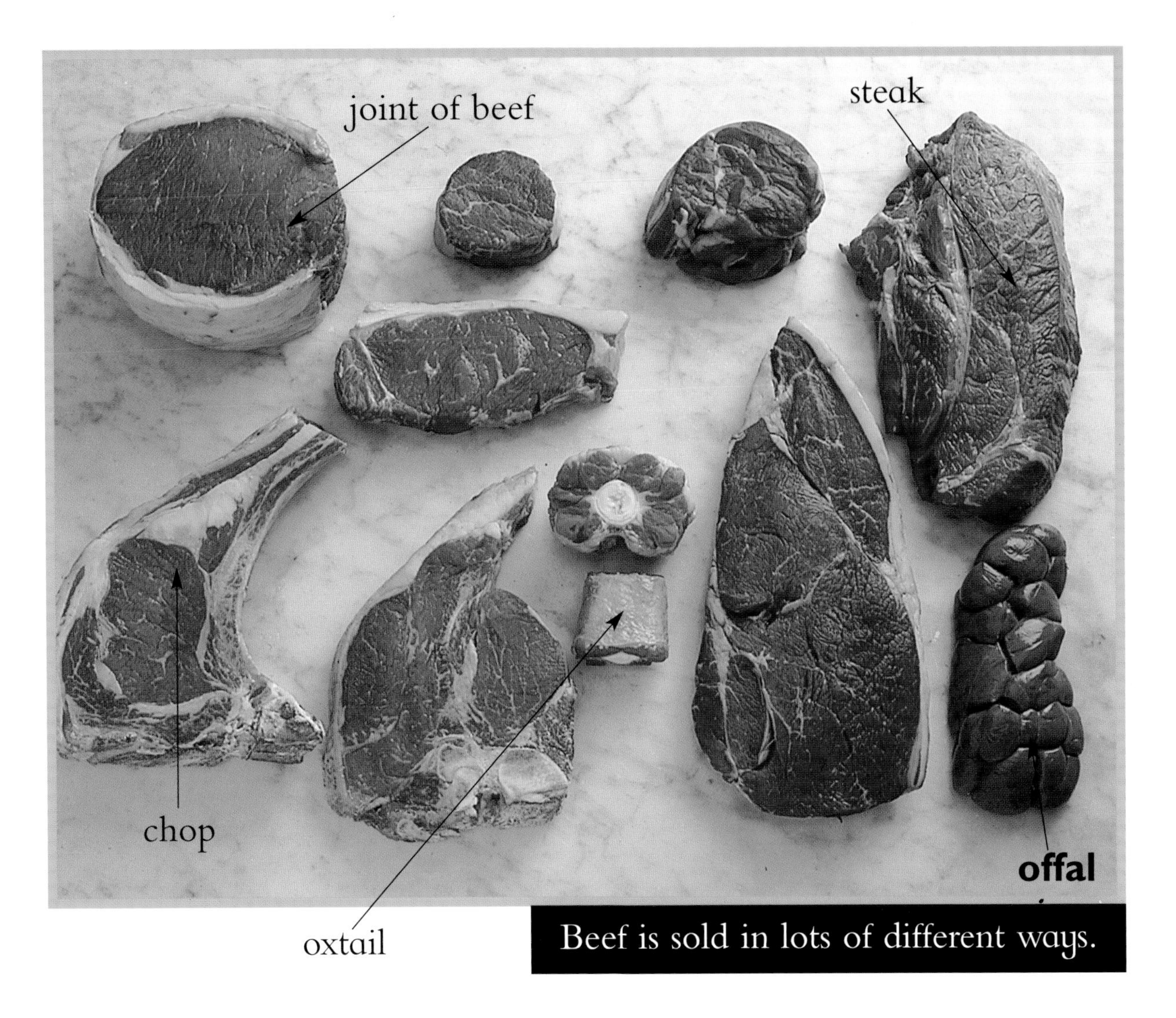

Beef is sold in lots of different ways.

Some farms keep cattle for their meat. This meat is called **beef**. It is sold in different ways, such as steak, mince or a joint of beef.

Other kinds of cattle farm

Cowboys round up these cattle on horseback.

In Australia and **the Americas**, **beef** cattle farms are very large. The cattle are almost wild and wander across many kilometres. **Stockmen** and **cowboys** round them up on horseback.

This road train is carrying cattle to market.

In Australia, cattle stations may be over 1000 kilometres from the nearest city. The cattle are carried to city markets in special large lorries called road trains.

Cattle in different countries

The farmer uses cows to pull the plough.

In some countries, cattle are used to pull heavy loads and farm equipment. In Asia, cows pull **ploughs** through flooded fields so the farmer can plant rice.

In India, people think cows are very special and must not be hurt.

Rodeos like this are very popular in America.

In America, some cows are kept for **rodeos** where **cowboys** see who is best. Each cowboy rides a horse and uses a **lassoo** to catch a cow and make it lie down.

Factfile

A **calf** can walk and drink its mother's milk almost as soon as it is born.

The cow helps her newborn calf to stand.

A cow knows her calf by its smell. A cow's sense of smell is better than her eyesight. So, you might often see a cow nose to nose with her calf.

A **dairy** cow can eat up to 70 kilograms of grass every day. It can eat its own weight in grass every week!

The heaviest bull that people have heard of was called Dunetto. He weighed 1740 kilograms. Work out how many children would equal that weight.

Most leather is made from the skin of cattle.

All these things are made from leather.

A cow's skin is called hide. Some people make rugs from the hide, with the hair still on.

Bison are related to cattle. **Herds** of bison can be found on the grasslands of Canada and North America. There were about 60 million bison on the **Great Plains** from Mexico to Canada. People used to hunt bison for sport and their hides.

Glossary

the Americas North, Central and South America

bed put fresh straw down on the floor for the cattle to lie on

beef meat from cattle. Cattle kept for their meat are called beef cattle.

calf/calves young cows or bulls. You say one calf, many calves.

chew the cud bring food back up into the mouth from the stomach, to chew it again

cowboys men in America who look after cattle

dairy cattle that are kept for their milk

farm-hand person who helps on the farm

female girl or mother animal

graze nibble or eat. Cattle usually graze on grass.

Great Plains huge areas of open grassland in the centre of America

herd group name for cattle

injection special medicine which is given through a needle

lassoo rope used to catch cows

maize crop that gives us sweetcorn or corn-on-the-cob

male boy or father animal

mineral lick block that animals lick to help them grow

offal parts of a cow such as heart, kidney, brain or stomach when it is used for food

pasture fields of grass for animals to graze

plough machine with large blades pulled through soil to turn it over

rodeo show where cowboys show how they ride and lassoo

silage winter food. It is made by cutting a crop and covering it with plastic. It goes sour, or pickled, and keeps well.

stockmen people who look after the animals (which are also called stock)

straw thick, dried stalks from crops

udder the part of a cow's body that is like a bag, which contains milk

tag metal or plastic label

vet doctor for animals

weaned gradually stopped from feeding on the mother's milk

yard enclosed area near buildings

More books to read

Encyclopedia of Animals of the World,
Dorling Kindersley, 1999

But Cows Can't Fly and Other Stories,
Marilyn Halvorson, Junior Gemini, 1993

Index

a b c d e f g h i j k l m n o p q r s t u v w x y z

Titles in the *Farm Animals* series include:

Hardback 0 431 13333 6

Hardback 0 431 13331 X

Hardback 0 431 13330 1

Hardback 0 431 13332 8

Find out about the other titles in this series on our website www.heinemann.co.uk/library